volume 4: DEVELOPMENT AND UNDERDEVELOPMENT 1945–1975	volume 5: THE GLOBAL COMMUNITY 1975–2000	volume 6: INTO THE 21ST CENTURY 2000–	chapter	topic
THE END OF PEASANT CIVILIZATION IN THE WESTERN WORLD	INDUSTRIALIZED AND MULTINATIONAL AGRICULTURE	THE PROBLEM OF SURVIVAL AND THE PROMISE OF TECHNOLOGY	1	Land, agriculture, and nutrition
OLD DISEASES IN THE THIRD WORLD AND NEW ACHIEVEMENTS IN MEDICINE	OVERPOPULATION, DEMOGRAPHIC DECLINE, AND NEW DISEASES	POPULATION, MEDICINE, AND ENVIRONMENT: A RATIONAL UTOPIA	2	Hygiene, medicine, and population
THE AGE OF PRIVACY: HOUSING, CONSUMER GOODS, COMFORT	MEGALOPOLISES IN THE THIRD WORLD; MULTIETHNICITY IN THE WEST	TOO MUCH AND TOO LITTLE: THE MISERY OF WEALTH	3	Living: environment and conditions
WHITE-COLLAR WORKERS, MANAGERS, AND LABOR PROTEST	AUTOMATION AND DECENTRALIZATION IN THE POST-FORD ERA	"THE END OF WORK" AND THE NEW SLAVERIES	4	Labor and production
NUCLEAR ENERGY: THE GREAT FEAR, THE UNCERTAIN HOPE	THE SEARCH FOR ALTERNATIVE ENERGY SOURCES	NEW FRONTIERS IN ENERGY	5	Raw materials and energy
THE PRODUCTION OF THE AUTOMOBILE	ELECTRONICS AND INFORMATION SCIENCE	WORK WITHOUT WALLS	6	Working: environment and conditions
MAN AND GOODS ON FOUR WHEELS	THE AIRPLANE IN MASS SOCIETY	FROM THE EARTH TO THE COSMOS: THE EXPLORATION OF SPACE	7	Transportation
EVERYDAY ENCHANTMENT: THE TELEVISION	THE INFORMATION AGE: COMPUTERS AND CELL PHONES	CYBERSPACE: THE WEB OF WEBS	8	Communication
PARALLEL ROADS: DEVELOPMENT AND UNDERDEVELOPMENT	THE COLLAPSE OF SOCIALISM AND THE RISE OF NEO-CONSERVATISM	THE MANY FACES OF GLOBALIZATION: LOCAL WARFARE AND THE GLOBAL COMMUNITY	9	Economics and politics
MOVEMENTS OF LIBERATION AND PROTEST: THE THIRD WORLD AND THE WEST	FEMINISM, ENVIRONMENTALISM, AND THE CULTURE OF UNIQUENESS	UNIVERSALISM AND FUNDAMENTALISM: THE NEW ABSOLUTISM	10	Social and political movements
CONSUMERISM AND CRITICISM OF THE CONSUMER SOCIETY	THE INDIVIDUAL AND THE COLLECTIVE	AFTER THE MODERN: ENVIRONMENTALISM, PACIFISM, AND BIOETHICS	11	Attitudes and cultures

THE ROAD TO GLOBALIZATION
Technology and Society Since 1800

In private and in public, at work or at play, in every stage of life, we live with technology. It becomes ever more present, and our perception of its artificiality fades through daily use. Within a very short time of their emergence, new possibilities seem to have been with us always, and the new almost immediately becomes indispensable. The choices that technology dictates and the paths that these choices take appear to be the only choices and paths possible—undeniable, unquestionable—and we perceive as natural the constructed world in which we live.

Despite the opportunities that technology affords us, and the promises that it makes constantly, we greet it with a general discomfort, an uneasiness that often does not reach the conscious level. But the manifestations of environmental crises can no longer be considered in isolation. The Westernization of the world marches in step with the widening—and already yawning—chasm between north and south, as well as with the emergence of aggressive localism. War seems to have resumed its role as a common tool in international confrontation. New diseases alarmingly outpace scientific discoveries, and biotechnologies and genetic experiments obscure the line between the human and the inhuman. The importance of the question of meaning has not been lessened by the decline of the sacred; but this question seems to find no place in the universal logic of growth that overcomes difference to guide governing bodies as representatives of economic and financial power.

A renewed uncritical faith in Progress on one hand and a demonization of "techno-science" on the other are often associated with a lack of context that comes of technology and with the dominance of a logic that neglects history. This logic can verify correctness in predetermined ways, but it does not comprehend the complexity of the greater process of change: it appreciates the present and the immediate future, but it cannot perceive itself as part of a larger historical evolution.

The first aim of a social and cultural history of the technology of the last two centuries, then, is to offer a careful and coherent study of the roads that have led to the development of modern culture. The basic objective of this series is the reconsideration of the innovative changes that have taken place and their diffusion over time, rather than a description of their first appearances. These innovative changes have marked and continue to determine our daily lives, the way we work, our relationships, and the points of view that contribute to global diversity.

It is important to recognize that our interpretations of the 19th and 20th centuries are centered on the men and women of the West, on their histories and cultures. This is undoubtedly a biased point of view, and it would be misguided to think that this partiality could be overcome by a simple updating of knowledge. The changing of a point of view that is rooted in history probably requires insight into processes that operate well beyond our perception. Perhaps the globalization that is underway, with its various worldwide effects, is establishing itself through precisely this mechanism: it is forcing a confrontation among lifestyles and different cultural models in new, absolute terms.

In general, the common historiography treats technological innovations only in brief digressions, glossaries, or chronologies of inventions and inventors; but we cannot fill in its gaps by constructing a separate history. Our realization of the economic, social, and cultural importance of industrialization, and our perception of the process as uninterrupted and ever more pervasive, have caused us to re-evaluate both the transformation itself and the new landscapes that industry has created—linking technology to economics and to politics, and systems of labor and production to culture and to social movements.

We can group as the Age of Technology the events that have been paving the road to the future for the last two centuries. Understanding the risks and the opportunities involved in so rapid a transformation of our world will require a change of mind and an updating of our culture—both of which are impossible without a broadening of knowledge and a renewal of historical consciousness.

2
PROGRESS AND THE EMPIRES
1850–1900

PIER PAOLO POGGIO
AND
CARLO SIMONI

ILLUSTRATED BY GIORGIO BACCHIN

English-language edition for North America © 2003 by Chelsea House Publishers.
All rights reserved.

All rights reserved. No part of this publication may be reproduced or transmitted in any form or by any means without the written permission of the publisher.

Chelsea House Publishers
1974 Sproul Road, Suite 400
Broomall, PA 19008-0914
www.chelseahouse.com

Preceding page: The zoopraxiscope (see chapter 11), the projector developed in the United States in 1880 by the English inventor Eadweard Muybridge. (Illustration by Paola Borgonzoni.)

Cross-section of two multistory boarding houses of the "boulevards" of central Paris, gutted and rebuilt by Haussmann (see Chapter 2). This illustration first appeared in Magasin Pittoresque in 1883.

Library of Congress Cataloging-in-Publication Data

Poggio, Pier Paolo.
The Road to globalization : technology and society since 1800 / text by Pier Paolo Poggio, Carlo Simoni ; illustrated by Giorgio Bacchin.
p. cm.
Includes index.
ISBN 0-7910-7092-1
1. Europe—Social conditions—Juvenile literature. [1. Europe—Social conditions—19th century. 2. Europe—History—1789-1900.] I. Simoni, Carlo. II. Bacchin, Giorgio, ill. III. Title.
HN373 .P64 2002
2002007848

© 2001 Editoriale Jaca Book spa, Milan
All rights reserved.

Original English translation by Karen D. Antonelli, Ph.D.

Printing and binding by
EuroLitho spa, Cesano Boscone, Milan, Italy

First Printing
1 3 5 7 9 8 6 4 2

INTRODUCTION

The second half of the 19th century is known as the Age of Progress, celebrated in Universal Expositions and nourished by the application of science to technology and industry. This volume deals with the exploits and the achievements of a faith in the future, involving scientists, entrepreneurs, politicians, and public opinion, that took root in industrial society. The forces of labor, which were organizing to fight for the recognition of workers' rights, embraced the idea that the entire world could be transformed and could benefit from industrial expansion. Industry had at its disposal a revolutionary new source of energy, electricity, which when combined with chemistry created substances and materials that did not occur naturally—and thus augmented the already increasing flow of consumer goods. There were few voices of dissension or concern. The resistance to modernness that had begun with the 19th century appeared to have been totally defeated.

The idea of the march of Progress was not an entirely abstract one. It solidified in the flourishing of the capital cities and in the intoxicating possibilities of life in the "Belle Époque." Threatening tensions were discernable only in the most disadvantaged strata of European society and were quickly repressed. As every industrial power created its own empire, the conflicts among the European states became concentrated in a race for the conquest of the world.

This print from the end of the 19th century shows the Wasson Manufacturing Company of Springfield, Massachusetts, a factory that produced railroad cars for the lines of the countryside.

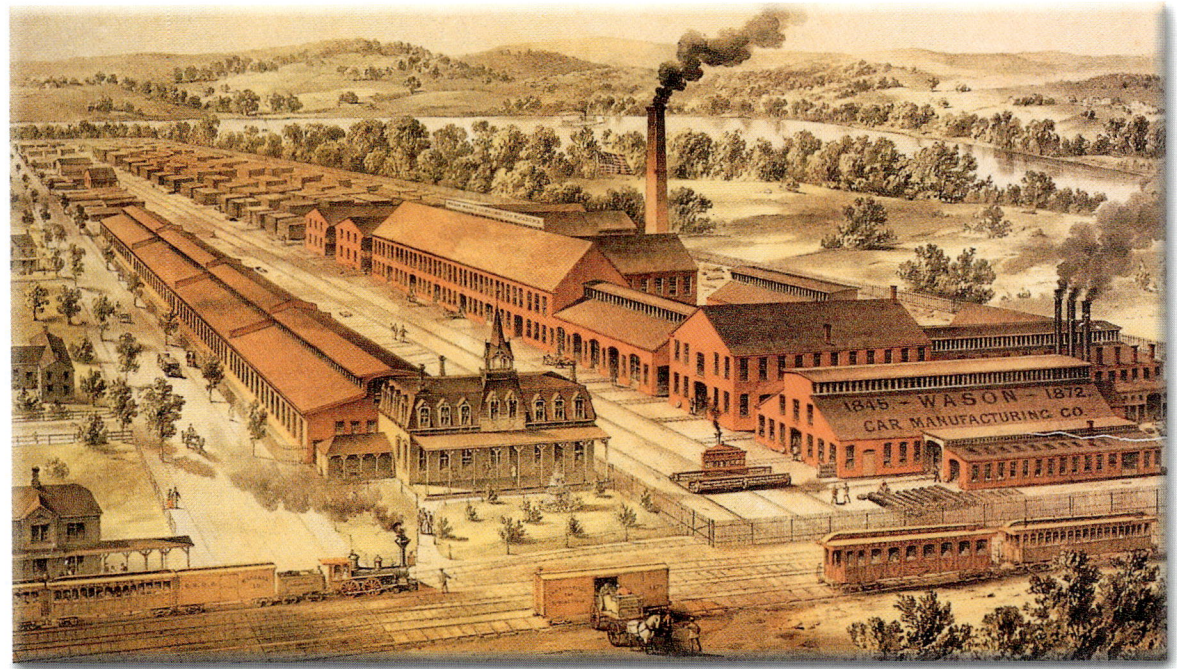

LAND, AGRICULTURE, AND NUTRITION

1. Plantation Agriculture and the World Agricultural Market

1. The planting of wheat on a grand scale on the Canadian plains, c. 1880. The slow settlement of Canada received a strong impetus in the last decades of the 19th century. During that time, the provinces of the prairies (Manitoba, Alberta, Saskatchewan) were systematically colonized.

▶ *Canada adopted practices of agricultural mechanization from the United States. The commercialization of hard (durum) wheat was made possible by the completion, in 1885, of the transcontinental Canadian Pacific Railroad and facilitated by the introduction of cylindrical grinders.*

Plantation agriculture, symbol of the white man's conquest, developed in the Americas before it did in Asia or in Africa. Plantations cultivated by slaves spread from Virginia to southern Brazil, though the slave-farmed sugar industry of northeastern Brazil had existed since the 16th century.

Merchants purchased slaves at favorable prices on the coasts of Africa. They sold them in America in exchange for products from the plantations—sugar, tobacco, cotton, coffee, and cacao—which they then sold at very high prices in European cities. This was a triangular commercial transaction that made it possible to amass vast fortunes. Even industrialization acted as a stimulus to plantation agriculture, for the development of the mechanized textile industry led to a spectacular increase in cotton plantations in the United States. By 1860, the eve of the Civil War, cotton by itself represented two-thirds of the American export.

In the rest of North America, the elimination of native populations and the absence of the remnants of feudalism allowed the establishment of an individualistic and capitalistic agriculture. In Europe, on the other hand, the agrarian individualism, and the elimination of community ties, was able to progress only despite many obstacles. Also, the crop yield in Europe remained fairly low in the 19th century.

American supremacy was the result of the union between agriculture and industry. Mechanization, in particular reaping and threshing machines, led to spectacular developments in production. In the United States, per-worker productivity tripled between 1840 and 1910. Long-distance commerce has been a powerful factor in economic and cultural growth throughout the history of civilization; in the last four decades of the 19th century, commerce grew as a result of technological innovations and the elimination, based on the English model, of customs duties.

It was only toward the end of the century that the European states

reimposed tariffs; they did this in order to slow down the competition resulting from foreign agricultural products. These products had become more readily available to European consumers because of the development of the steam ship, the expansion of the railroads, and the invention of refrigeration. In the second half of the 19th century, large quantities of grain at greatly reduced prices arrived in Europe from the Americas, Australia, and India. The exportation of meat from the United States to Europe increased spectacularly. The same was true for Argentina, Australia, and New Zealand. By the end of the century, a global agricultural market had been formed, varied by regional specializations. For the exporters of raw materials, however, who specialized through a system of plantations and single-crop production, much depended on the trends of international commerce, which in turn were in the hands of a few political and economic forces concentrated in the nerve centers of the international economy.

2

3

2. Jean-François Millet, Gleaners, 1857, the Louvre, Paris. The mechanization and the industrialization of agriculture progressed fairly slowly in 19th-century Europe. The causes are linked to social, demographic, and environmental factors more than to technological factors or to the mentality of the farmers. The structure of the land holdings and the abundance of predominantly female labor favored traditionalism.
3. The backwardness of agriculture on the old European continent favored the products that came from North America. Of particular interest were the inexpensive foodstuffs that came from Central and South America.
4. A cattle farm. The raising of cattle on a grand scale spread after the colonization of Latin America. Due to the perfecting of transportation and the introduction of new refrigeration techniques, the pampas of Argentina, Brazil, and Uruguay gained global importance as a center of meat production.
5. Work in a Cuban sugar factory. Cuba's dependence on the cultivation and production of sugarcane dates to its first productive expansion, from 500 to 10,000 tons, in the 18th century. Sugar and slavers were the pillars of Cuban society well into the 19th century. Independence from Spain, obtained at the end of the century, depended on the influence of the United States, which held a monopoly on the sugar market.

HYGIENE, MEDICINE, AND POPULATION

2. THE RECLAIMING OF THE CITIES AND THE DISCOVERY OF MICROBES

1. The leveling of the urban areas in front of the Paris opera house. This restructuring was ordered by Napoleon III. The results of this intervention were many, but without a doubt one was to build for the city of Paris a reputation as the most modern European capital. The plan isolated monuments, gutted old neighborhoods, and, championing the boulevard, superimposed wide streets upon the pre-existing ones to enable the easy circulation of traffic. Baron Georges-Eugène Haussmann (1809–1891) was the executor of the renovation of the French capital, which represented a boon for the economic and speculative interests of the time. Concurrently, the renovation established greater control over working-class neighborhoods, which already had been placed in the suburbs. In his plans, however, Haussmann had not shown much interest in the reorganization of these working-class neighborhoods.
2. The French biologist Louis Pasteur (1822–1895) is remembered for several fundamental discoveries, including the rabies vaccine. He differs from Jenner, who found his vaccine in nature, in non-virulent bovine smallpox germs, in that he developed immunology scientifically. He understood that it was necessary to reduce the virulence of the virus before injecting it so that it would not attack the organism and would only stimulate the production of antibodies, which would then lie in wait, ready to strike on contraction of the disease.
3. Inmates in a sanitarium at the end of the 19th century. Beginning with the end of the 19th century, to aid in the fight against tuberculosis, doctors prescribed sojourns in the mountains, or at spas where the weather was mild, with increasing frequency.
4. Robert Koch (1843–1910). The German bacteriologist responsible for the introduction of the technique of cultivating bacteria outside a living organism used a serum that he kept at body temperature. His fame, however, is due to the discovery of the tuberculosis bacterium. The announcement of the discovery was made in Berlin in 1882.

The cholera epidemic that struck Europe in 1832 indicates that certain diseases had not been defeated in the poor neighborhoods of the cities and that the urban environmental degradation was a general problem.

The belief that the spreading of the illness from person to person was not based on contagion but on the influence of ill-defined "miasmas," or stenches present in the air, led to the decision to tear down the walls that surrounded cities and to open up ancient neighborhoods to permit the circulation of air. Much sounder reasoning led to the building of sewer systems in London and Paris in the 1850s to prevent the contamination of drinking water.

The works within the cities to restore healthful conditions were a decisive way to confront epidemic diseases, like cholera, and other gastroenteritic maladies that were responsible for the high rate of infant mortality. Only developments in medicine, however, were able to combat scourges like smallpox. Mass vaccinations in the first decades of the 19th century led to the near-disappearance of smallpox in Great Britain and to a decrease in its incidence in other European countries.

Then, in the second half of the century, Pasteur and Koch conducted studies that demonstrated the relationship

between disease and bacteria. Their discoveries had many consequences, even to daily activities; for example, hand washing increasingly became perceived as a social obligation.

Great advances in immunology were accompanied by innovations like the stethoscope and blood pressure machines and the use of the thermometer, as well as other significant discoveries, from the recognition of the importance of hygiene and the resultant disinfection of operating rooms to the power of anesthetics (at first morphine and then ether and chloroform) to eliminate pain during surgery. Therefore, it was no great surprise that the social prestige of medical doctors rose with the public's faith in them. The 19th century is considered the "heroic century" for medicine.

5. *As early as the 19th century, in addition to problems of overpopulation and hygiene, the problem of traffic congestion arose. In 1838, the number of horse-drawn omnibuses in London had increased to such an extent that the city considered the construction of an underground steam-driven railroad. The first segment of the London Underground was inaugurated in 1863.*
6. *The famous French medical doctor Jacques Rohr painted his* Scene of Radiography *in 1886. The technique was used ten years later in experiments conducted by Wilhelm Conrad Roentgen.*
7. *Robert Hinckley's 1882 rendering of a public demonstration of surgery under anesthesia that had been performed successfully in 1846 at Massachusetts General Hospital. Following the use of "laughing gas" (nitrous oxide), and morphine and ether, it was chloroform that opened the door to modern anesthetic technique.*

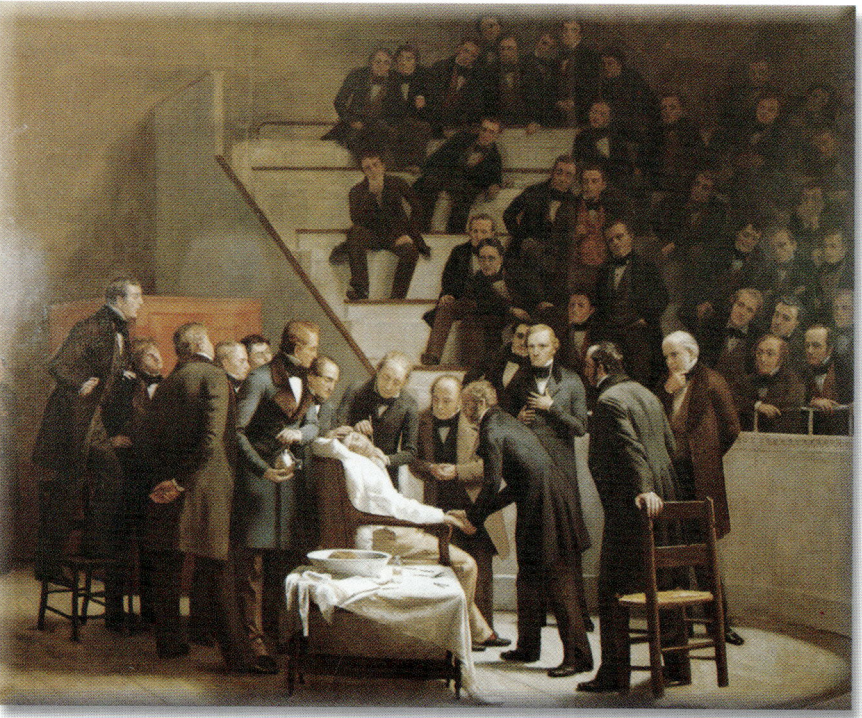

LIVING: ENVIRONMENT AND CONDITIONS

3. European Capital Cities and the Promised Land Overseas

At the beginning of the 19th century, no European city had a population of one million inhabitants. Within a few decades, the situation changed radically.

Either directly or indirectly, the Industrial Revolution was at the origin of this phenomenal urban growth. True factory-cities were born, and ports became more active; and within these cities developed a concentration of commerce and consumers. Concurrently, administrative activities received a new impetus. Capital cities like London and Paris were at the forefront of these developments, concentrating within their boundaries the functions of politics, administration, economics, and culture.

A typical example of the rapid development of an ancient capital is Vienna, whose population increased from 440,000 inhabitants in 1850 to 1.1 million in 1880. Berlin's population increased from 172,000 inhabitants in 1800 to more than one million in 1880 and double that number by 1910. The expansion of Paris outpaced these: from half a million inhabitants in 1800 to nearly 2.5 million in 1880. London's great size reflected British hegemony during the 19th century: shortly after the beginning of the century, there were slightly more than one million residents living in the city; by 1850, the number had risen to 2,685,000; in 1880, there were 4,770,000 people; and by 1910 the population had swelled to 7,256,000 inhabitants.

Even the unrestrained urban development of the time could not accommodate the waves of peasants who, unable to meet the ever-increasing competition from world agricultural markets, were migrating from the countryside. As a result, emigration overseas increased, and by the middle of the 19th century emigration from Europe to America was considerable. There were many causes of this displacement: the famine in Ireland; the discovery of gold in California; the development of steamships and railroads; agricultural overpopulation. Humanity, even though characterized by frequent population movements, had never before recorded a migration of such proportions.

Between 1841 and 1880, approximately 13 million Europeans left the old continent. A similar number emigrated in the century's final decades. Departures increased even in the following years, reaching two million in 1910. Even taking into account repatriation, an emigrant's return to his native land, emigration modified the demographic composition of the countries of origin as well as those of destination. These changes affected relationships among the sexes and the generations, and emigration played a decisive role in the diffusion of cultures and European lifestyles. Specifically, it contributed to the establishment of the United States as the leader in industrial development.

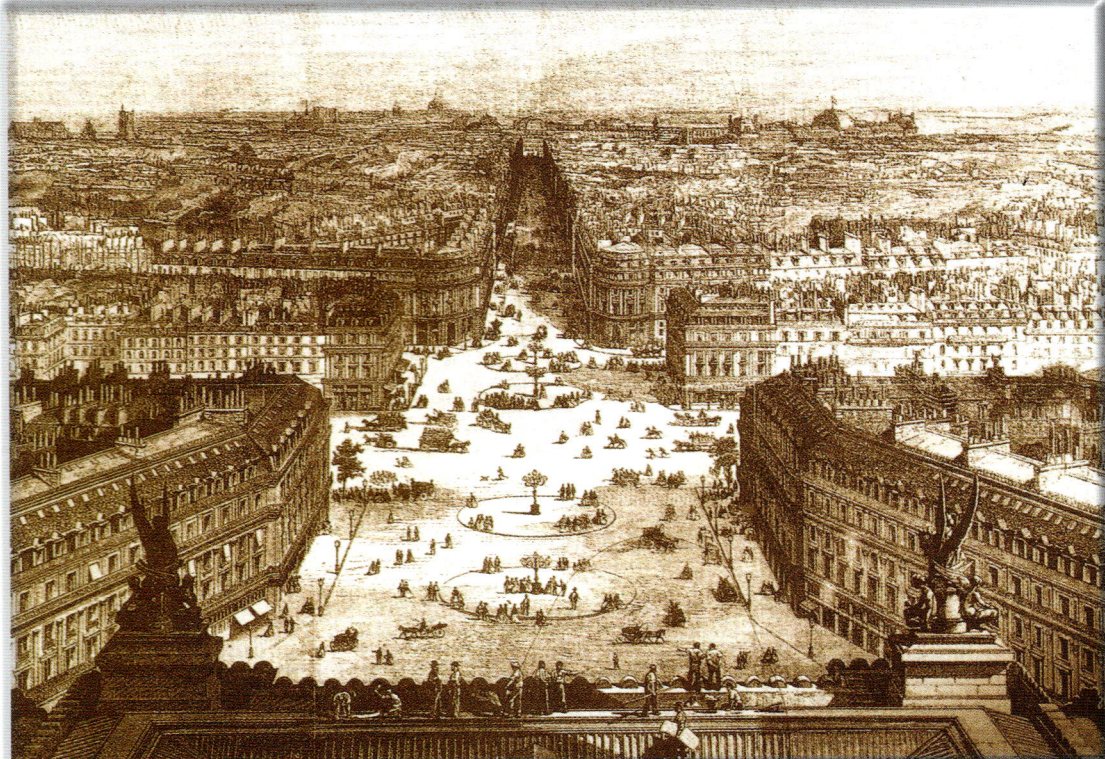

1. Paris in the Second Empire. Under orders from Napoleon III, Baron Georges-Eugène Haussmann, the Parisian prefect, began the transformation of the city in 1853. The master plan called for the gutting of the old city and the construction of a system of great boulevards, wide sidewalks to serve the shops and the buyers, and a multiplicity of passageways and the opening of department stores.
2. Edouard Manet, Music in the Tuileries (1860–1862), National Gallery, London. As did Manet's other works, this painting caused a scandal among traditionalists. It expresses the artist's support of the modernization of customs through the diffusion of worldly rituals—on the stage of the bourgeois capital of art and culture.

3. Arrival in New York was one of the symbols of the migration from Europe to America.
4. A family of Italian emigrants arriving in New York; photograph by L. Nole, 1905. At the end of the 19th century and the beginning of the 20th century, Italian emigration focused on the Americas. At the outset, the migratory flow headed mainly toward South America, but by the 20th century the more desirable destination was the United States, which at that time had blocked Asian emigration legislatively.
5. A medical examination on arrival at Long Island, New York, 1905. The most intense period of transatlantic emigration was from the beginning of the century to 1913 and was characterized by an intense back-and-forth movement. The majority of the emigrants were ready to return home as soon as they were able and saved as much money as possible to send to their families.

LABOR AND PRODUCTION

4. Skilled Labor and the Organization of Research

From the point of view of production, the most effective innovations of the first stages of the Industrial Revolution were the accomplishments of the artisans, the specialized workers, and the self-taught mechanics. The same is true of the cotton textile sector with James Hargreaves, Richard Arkwright, and Samuel Crompton. The innovations and improvements made to machine tools were also the result of the extraordinary tradition of the machine-operating artisans that characterized England. This equipment included Wilkinson's cylinder boring machine (boring mill), Maudslay's lathe, and Whitworth's precise machine tools.

In addition to the actual inventions, we must keep in mind the adaptations and the small improvements carried out in the factories themselves. These resulted from the direct application of the workers' knowledge to the job. The specialized workers, who were coveted and very mobile, were the principal channel by which spread the technological innovations that were utilized creatively in local production. On the other hand, in many industries, ancient artisan traditions and trade techniques were preserved within the new labor and production organizations.

Great Britain, although liberal regarding commerce in products, was extremely protective of its production secrets. The countries of the Continent were forced to focus on technical instruction and on applied research, from the École Polytechnique in Paris to the German Technische Hochschulen; the first German polytechnical academy, was founded in Karlsruhe in 1825.

Through the creation of specialized institutes and laboratories tied to industry and technological and scientific research, Germany was able to identify ways to overcome the gap that separated it from England. Among the research projects that led to successive practical applications, the following were the most important: Liebig's opening the path to chemical fertilizers; von Hofman and Kekulé's enabling the production of synthetic dyes and, successively, a large quantity of other products; and Diesel's inauguration of the diesel engine at the end of the century.

The second Industrial Revolution confirmed the directly productive role of scientific research, especially in the sectors of chemistry, electricity, and the automated production of machines. Scientific research became the highway

1. Ferdinand-Joseph Gueldry, The Grinders, 1885, Museum of Art and Industry, Saint-Étienne, France. Work that was performed by hand, like the grinding of mirrors, existed next to the development of factory systems and large industries.

2. Workers and apprentices posing in an artisan shop. Contrary to popular belief, industrialization did not mark the demise of the artisan. Old trades resurfaced, and new trades emerged. The artisan shop was the venue for learning important trade secrets, for building refined devices with basic tools, and for acquiring professional awareness and pride.

of development that would determine the leaders in the race among nations in the following century.

Under 19th-century conditions, the English approach, imprinted with the traditional empiricism of Anglo-Saxon culture, best guaranteed success. This pivoted on the knowledge of artisans and technicians and was affirmed by the initiative of entrepreneurs. Similar events occurred in the United States, where technology was capable of reducing the labor force and where standardization and mass production held privileged roles.

3. Liebig laboratory in Giessen, 1842. Justus von Liebig (1803–1873), through his studies in organic chemistry and on the nutrition of plants, made a fundamental contribution to the production of chemical fertilizers. His laboratory exemplified the rapport between technology and the German university. Ferdinand Redtenbacher's science of machine construction in Karlsruhe provides another symbolic case.
4. Hand-milling of mass-produced pieces. Within a few decades, specialized workers who operated machine tools had to demonstrate professional ability with a new method of production, based on the use of electrical energy applied to implements of steel.
5. A large hand-operated vertical drill. Even if operated by hand, the machine tools were now built entirely of metal and had mechanical advancements and precise settings. The problem facing the 1870s was to provide the energy necessary to power the machines operated by artisan mechanics.

RAW MATERIALS AND ENERGY

5. ELECTRICITY: SCIENCE, INDUSTRY, AND IMAGINATION

1

1. A lamplighter, a common figure in 19th-century cities illuminated by gas lamps. At first, the new form of energy was used only in factories. Eventually, it was used in public areas, and finally, not without arousing fears similar to those that had accompanied steam, the new form of lighting was used in private homes.

Even in ancient times, electricity was seen as a powerful and mysterious presence. In the 18th century, it was studied by numerous scientists in its natural manifestations, such as lightning. In 1800, electricity became the object of experimentation when Alessandro Volta's cell enabled the production of a continuous flow of electrical current. Twenty years later, the Danish scientist Hans Christian Oertsed discovered the link between electricity and magnetism, noting that a metal wire charged with electric current caused a nearby compass needle to move.

Oertsed's discovery, which was studied in greater depth by André-Marie Ampère, led to the invention of the electromagnet and its application to the telegraph, to mechanical electrical generators (dynamos and alternators), and to the first electric motor, which had been attempted by Michael Faraday as early as 1821.

However, the use of electricity that most profoundly changed daily life was in lighting. In 1813, the year in which he created the safety lamp for miners, Humphry Davy held a demonstration of his new arc lamp in London. It was not until 30 years later that the arc lamp was tested to light public areas in Paris and, subsequently, in London and several cities in the United States.

The harsh light emitted by the arc lamp, and its unreliability, did not favor its use within the home. The change came with the incandescent bulb, invented by Joseph Swan in 1878 but perfected and made truly efficient in the following year by Thomas Alva Edison. Thus electric lighting came into the home. Even the most modest homes had electric lighting, thanks both to Edison, who progressively lowered the cost of light bulbs, and to the elimination of the fear of explosion that had haunted the use of gas lighting.

Through the use of the steam engine, which had begun during the first Industrial Revolution, inventors were able to produce large quantities of energy to run central turbines and thereby produce electricity. The result was the birth of a new type of large-scale production requiring enormous concentrations of capital. This in turn became a key component of what would be known as the Second Industrial Revolution, which was characterized by a myriad of inventions and devices that were made possible only by the new form of energy. These creations have become so integral a part of modern daily life that they appear as natural fixtures in our surroundings.

16

2. A German city lit by arc lamps in the 1870s. Although an object of wonderment and enthusiasm, this invention represented in some respects a step backward from gas lighting. The intensity of arc lamps could not be adjusted; nor could a common system of supply be established in order to feed more than one light. (An image from an 1873 publication from Leipzig.)

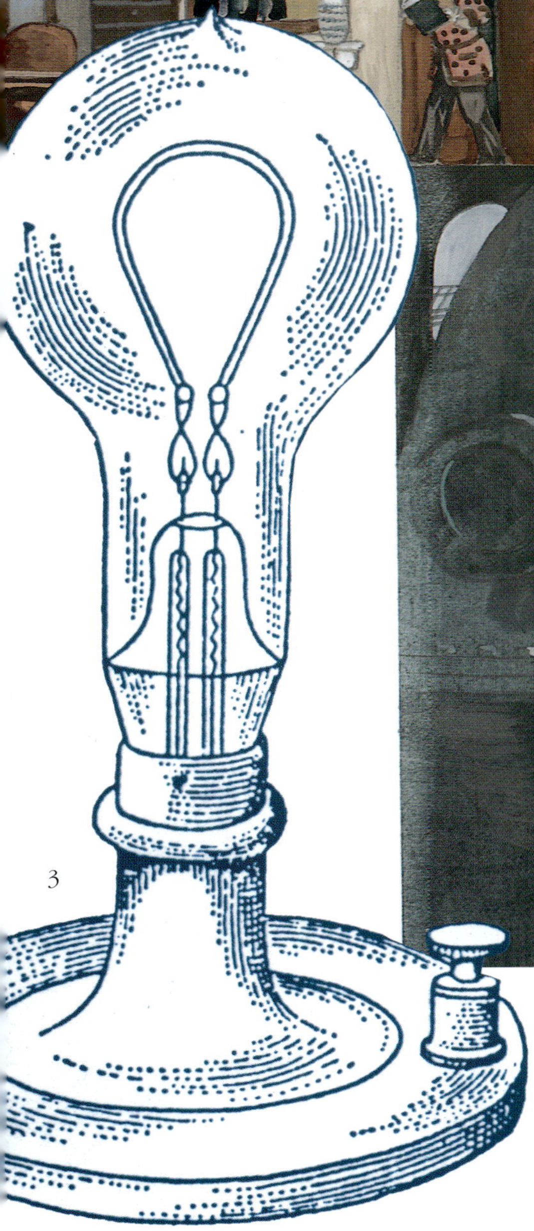

3. Edison's bulb with a carbon filament (ca. 1881).

4. Electricity radically changed daily life, even in the private sphere. The atmosphere and the habits within the home were altered so drastically as to render inconceivable any living accommodation without the new comforts.

5. Large turbines installed in the main hall of an electric plant from the end of the 19th century. The newness of the energy source and the technologies used in its production in many cases stimulated the creation of sumptuous and refined structures.

WORKING: ENVIRONMENT AND CONDITIONS

6. THE "MYTH" OF STEEL AND THE REALITY OF THE FACTORY

The mythic nature of steel, linked to power and to war, has a history that begins far back in time. In the age of the Industrial Revolution, the "myth" acquired new strength; through the use of steel, machines were built that enabled man to transform matter and to defeat his enemies.

Despite the textile industry's greater importance from an economic and social standpoint, during the second half of the 19th century the symbol of industrial success was the production of steel. For many years to come, the power of a nation was measured in tons of steel.

For the first three quarters of the century, the English metallurgical industry maintained its dominance because of the availability of coke which stoked its blast furnaces. Successively, in what was to be known as the Age of Steel, the situation changed, due to the innovative genius of amateurs like Henry Bessemer and Sidney G. Thomas. Around the middle of the 19th century, two significant changes took place in the metallurgic sector: the definitive triumph of combustible minerals over charcoal, and the substitution of iron with steel, thanks to Bessemer's converter (1856) and Martin-Siemens' furnace (1864). These changes coincided with the rapid expansion of the European and American iron and steel industries.

In 1870, the United Kingdom produced half of the world's cast iron, a quantity which was 3.5 times the production in the United States. In 1890, however, the United States permanently took the lead in the production of both iron and steel.

Toward the end of the century, Germany too was able to surpass England's production. The demands imposed by the railroad industry, which was in a period of full expansion, determined the rapid growth of the iron and steel industries. The iron and steel workers represented a kind of aristocracy among workers because of the exceptional physical strength their jobs required, the risks of the process, and the empirical knowledge they had to acquire and apply. They worked in extreme conditions, and their life expectancy was very short. Theirs was not an isolated case, however; risk, fatigue, and disease were constant features of factory and mine work. In the second half of the century, partly due to an economic boom, the actual wages of workers improved, but this was not the case for their working and living conditions. The work hours remained extremely long, and the introduction of gas lighting allowed the general implementation of night work. The speed of the work was dictated by the machines, and the workers' movements were overseen and subjected to discipline all the more.

The great riches that resulted barely touched the workers, male or female. Child labor was the norm. At this time, female labor was extremely important, beginning with the textile sector. Women in the factories were forced to perform the most menial, repetitive, and tiring tasks in unhealthful and uncomfortable environments. In addition to their factory jobs, they had to care for the home and family.

1. Gustave Eiffel, Viaduct across the Viaur River, Albi, France, 1897–1902. In addition to the famous Eiffel Tower, erected for the World's Fair of 1889, the engineer Eiffel (1852–1923) built large iron structures in numerous countries in the world. Of particular note were the viaducts that enabled Eiffel to pair his own innovations with a technology that had become symbolic of progress.
2. Poster advertising products manufactured from steel. The publicizing of prefabricated steel products was the inspiration of Daniel D. Badger (1806–1884). Badger was among the most active proponents of the use of steel in architecture—to the point of advocating the construction of towers tens of kilometers in height.

3. *A skyscraper under construction. With girders weighing thousands of pounds, steel was the key to a complex technological machine that arose as the symbol of American modernness. On the other hand, buildings constructed of metal were restricted to the outskirts of the ancient European cities. In any case, the skyscraper is one of the few inventions of modern architecture.*

4. *Louis Henry Sullivan, Buffalo Guaranty Building, Buffalo, New York, 1894–1895. This skyscraper, planned with D. Adler, is considered one of the masterpieces of contemporary architecture. Sullivan did not use the steel framework as it had been normally used, to construct a purely utilitarian neutral object; he attempted to give life to the skyscrapers, the new protagonists of the nascent metropolitan landscape.*

5. *The Krupp cannon, with a 1,000-pound breech. This weapon, displayed at the Paris Exposition of 1867, was used in 1870 by the Prussians to bombard the same city. The company founded by Alfred Krupp (1812–1877) became famous for the invention of the steel cannon and was among one of the first to use the Bessemer process.*

6. *Adolf von Menzel, The Iron Rolling Mill, 1875, Staatliche (State) Museums Berlin. Menzel was particularly close to the working class and participated in the 1848 revolution. He was also full of enthusiasm "toward the Cyclopean world of modern engineering." In his paintings he tried to represent the factory using strong and effective tones.*

7. *Thomas Pollock Anshutz, The Ironworkers' Noontime, 1880, San Francisco Fine Arts Museum. The artist had lived in the steel town of Wheeling, West Virginia and was well acquainted with the world of the workers. In this painting, he avoided the use of high and rhetorical tones, giving us an image of daily factory life.*

TRANSPORTATION
7. LIGHTNESS AND STRENGTH: FROM THE SAIL TO STEAM

Beginning in the 17th century, the fast and efficient European three-masted sailing ship, a versatile instrument of commerce and of war and the product of a century of innovations and improvements, became the means by which Europe imposed its domination.

The sailing ships had formed the basis of knowledge for builders and sailors and had inspired flights of imagination that later would be immortalized in the stories of Conrad and Melville.

Keeping these aspects in mind, as a result, it comes as no surprise that the substitution of the power of steam for the force of the wind did not come into acceptance and common use for decades. The steam-powered ship was essayed first on a French river, in 1783, and later at sea, in the first years of the 19th century.

There were several factors that discouraged this new form of propulsion: the lack of power and the uncertainty of the first steam engines; their weight, which required large iron structures; and the sacrifice of storage space to the necessary large quantities of coal.

Even in the second half of the century, large sailing vessels continued to be built, and sails remained on the decks of those ships equipped with wheels or propellers powered by steam. However, this artificial form of energy had by now demonstrated its ability to overcome the natural obstacles of wind and waves on the sea—even before overcoming the ruggedness of the land—allowing movement without dependence on environmental conditions.

In 1824, Sadi Carnot, the father of thermodynamics, stated in his *Reflections on the Motive Power of Fire* that "[s]team-powered navigation brings the farthest countries closer together, it unites the peoples of the earth as if they lived in the same country." "Were not the shortening of the duration, the lessening of the hardships, and the dangers of the voyages equivalent in practice," he asked, "to a noticeable shortening of distances?"

In truth, the importance of steam navigation was established only when it entered into competition with the sailing vessels that could not rely on regular winds for their specific routes—through the Suez canal, for example, which opened in 1869.

Another determining role was played by the use of steel instead of iron. This change lightened the ships while still guaranteeing the capacity to accommodate large loads of coal. The other technological addition at the end of the century was the introduction of powerful steam-driven turbines.

1. In 1818, the American steamship Savannah was the first to use an auxiliary steam engine during an ocean crossing. The available space below deck was almost entirely filled by the engine, the boiler, and the coal.
2. In 1845, the English gunboat Alecto became the protagonist of a historic tug-of-war when it was pitted against another vessel, the Rattler. Both boats had the same characteristics, but the Rattler was powered by a propeller and the Alecto by a paddle-wheel. The Rattler and the Alecto, connected stern to stern by a cable, pushed their machines to full force. The Rattler demonstrated its superiority; it soon was towing the Alecto.
3. The English steamer Sirius is considered the first completely steam-powered ship to cross the Atlantic Ocean—from London to New York, in 1838.

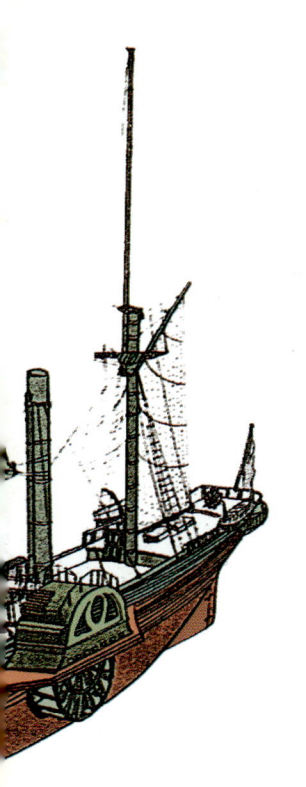

4. The contemporary imagination was struck by the unusual dimensions of the transatlantic steamers. Jules Verne was inspired by the British ship Great Eastern *(launched in 1858)* in the writing of his novel A Floating City. The French writer said that the transatlantic liner was a "microcosm which brought the world with it."

5. *Isambard Kingdom Brunel, originally a railroad engineer, dedicated himself with passion and inventiveness to the construction of ever larger and more powerful ships. He ended his career with the launching of the transatlantic liner* Great Eastern. *This ship was used a few years later to lay telegraph cable across the ocean floor.*

6. *Joseph Mallord William Turner,* The Fighting Temeraire, Tugged to Her Last Berth to Be Broken Up, 1839, *Tate Gallery, London. One of the first depictions of the historic confrontation between steamships, it represents in the foreground the small tugboat and colossal sailing vessels.*

7. *The* Mauretania, *launched in 1906, could carry 3,000 persons. The small masts at the stem and bow, practically without function, were kept as testimonials to the expired epoch of the sail. Ocean liners reflected the social differences of the time in their rigid divisions between decks and classes, luxurious cabins for the wealthy and huge common rooms for the emigrants, ballrooms and machine rooms. For these reasons, the ocean liner became the backdrop for stories and films set during the first decades of the 20th century.*

COMMUNICATIONS

8. THE SLOW PROGRESS OF THE TELEPHONE

"A telegraph that can transmit the human voice" was how Alexander Graham Bell defined his invention in 1876. In addition to the functions that tied it to the telegraph, in reality the telephone had introduced a basic innovation: the possibility of not merely transmitting a signal, but reproducing from afar a sound with all its different modulations.

The telephone companies that prospered in the United States in the last two decades of the 19th century seem not to have been aware of the scope of this innovation. For example, AT&T, which held Bell's patent, was headed by managers who had grown up in the world of the telegraph and therefore leaned toward considering the new instrument as an improvement on the old, to be used in the same fashion—to communicate economic and financial information, political and military dispositions, or, on occasion, emergencies. For these reasons, the telephone companies, immersed in a commercial and industrial circuit, connected businesses before private homes, and urban families before rural ones. At the same time, however, a new social use was being defined for the telephone, predominantly by American housewives. These women used the instrument to maintain contact with the world outside the family circle. Through telephone conversations they integrated the fundamental task of maintaining relationships with relatives, neighbors, and friends that made up part of their responsibilities of nurturing and care.

In Europe, the telephone encountered a different resistance. The new instrument widened unimaginably the sphere of communication, but it confined communication to the present, forcing it into an unusual timeliness. This was different from a letter, which left room for meditation both for the writer and the recipient; the letter was an instrument of preserving the past. The intrusiveness of the telephone ring disturbed many people of that period, who regarded it as a violation of their private space and of their personal time.

1. A stanchion telephone from the end of the 19th century. The shape of the instrument, which was equipped with two distinct microphones, one to transmit and one to receive, was based on the conviction that the power of the magnet contained in the apparatus was proportional to its length.
2. Alexander Graham Bell (1847–1922) during his first call on the New York–Chicago line.
3. The October 6, 1877 issue of Scientific American *dedicated its cover to Bell's New York–Chicago call.*

These feelings were not limited to the common people. As early as 1877, Bismarck connected his Berlin office with the farm where he passed most of his time, 230 miles away. In contrast, the Austrian Emperor Franz Joseph forbade the installation of telephones at the Hofburg, the historic imperial Hapsburg residence in Vienna.

4. *Until Almon B. Strowger invented automatic switching, in 1899, telephone communication required a great number of operators, usually recruited among the female population, who connected cables on a control panel.*
5. *Social satire of the time dealt at length with what it considered to be an abuse, or at least a distortion, of the function of the telephone—for example, its use in amorous conversations.*

ECONOMICS AND POLITICS

9. THE TRIUMPH OF IMPERIALISM

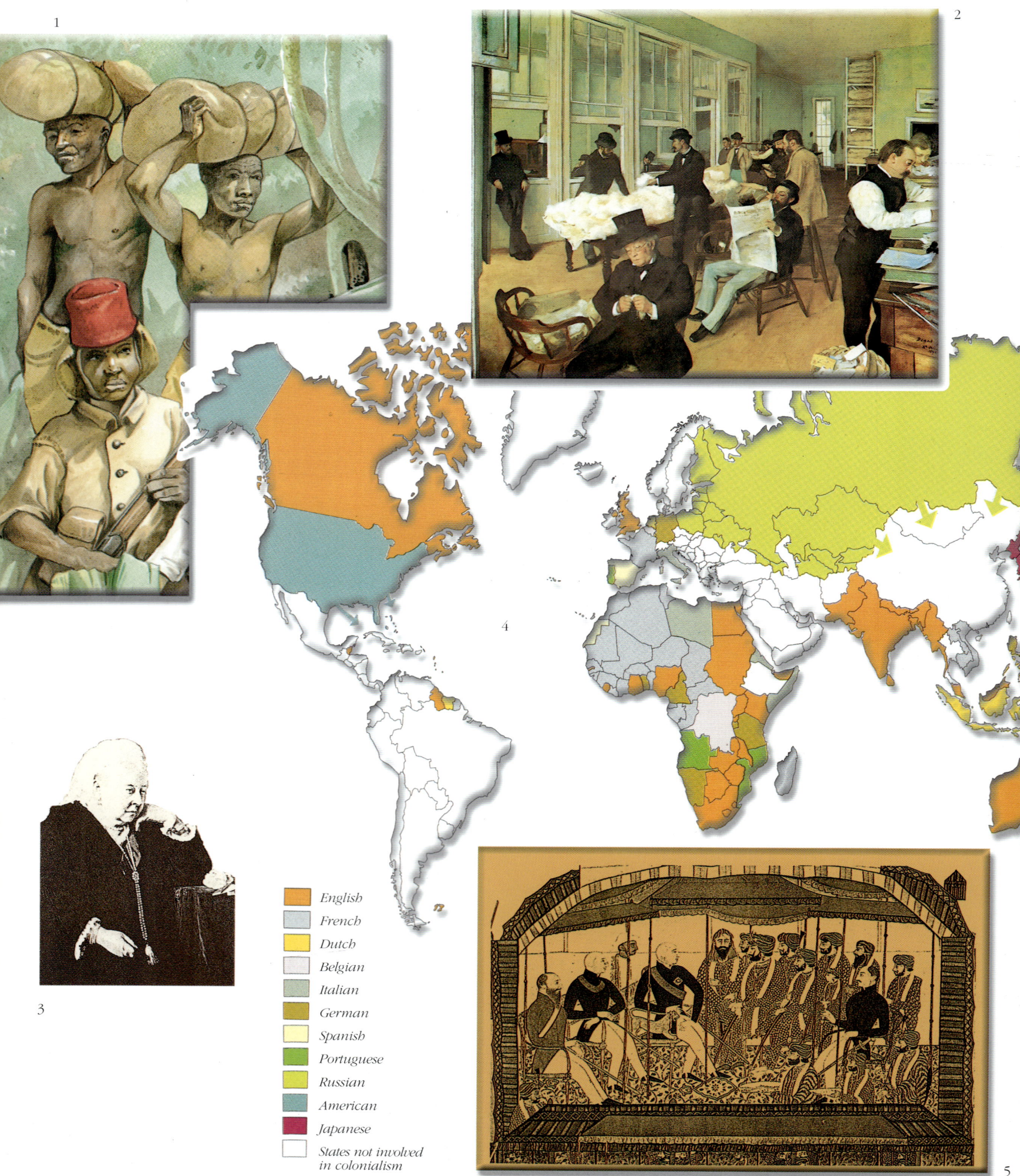

English
French
Dutch
Belgian
Italian
German
Spanish
Portuguese
Russian
American
Japanese
States not involved in colonialism

4. The division of the world by colonial powers between 1870 and 1912. The Second Industrial Revolution, called the Age of Imperialism, made possible, and intertwined with, an intense cycle of commercial and colonial expansion. Searching for markets and raw materials, the European nations conquered the world in the name of white civilization.

5. A meeting between Indian and British representatives during negotiations over Punjab. In 1849, Punjab was annexed to Great Britain. England had controlled India indirectly for a long time through the British East India Company, but now it aimed at direct domination. A turn of events occurred with the Great Revolt of 1857–1858 and its subsequent repression; India became an official colony of the United Kingdom.

6. The Boxers attack the foreign neighborhoods of Tientsin in this poster from the British Museum, London. After the Opium War of 1840–1842, the European powers intensified their penetration into China, which had been shaken by insurrections. The Boxers gave voice to the strong sentiment against the "foreign devils." The defeat of the Boxers permitted the colonial powers to impose an "open-door policy" (freedom of economic penetration).

1. Colonialism in Africa. With the exception of British South Africa and French Algeria, direct colonization of Africa began only after 1880. Penetration was facilitated by the disintegration of the Ottoman Empire. In 1882, the English established Egypt as a protectorate. After the Congo was won by Belgium in 1885, a race for the division of Africa developed among the European states. (Sketch by A. Baldanzi.)

2. Edgar Degas, Portraits in a New Orleans Cotton Office, *1873, Musée des Beaux-Arts, Pau, France. Degas (1834–1917), the son of a banker and a Creole woman from New Orleans, painted this work after returning from a trip to America. Following the defeat of the South in the Civil War, the cotton monopoly of the slave states was weakened, but the port of New Orleans maintained its importance.*

3. Queen Victoria (1819–1901). She succeeded to the throne in 1837 and at first governed under the influence of her husband, Prince Albert of Saxe-Coburg-Gotha, a strong supporter of technological and economic progress. After her husband's death, she upheld with conviction the imperialistic politics of Benjamin Disraeli, who wanted to see her Empress of India in 1877. By the end of her reign, the British colonies covered one quarter of the earth's surface.

The 19th century witnessed a spectacular increase in European influence in the world. Between 1800 and 1900, the area of the territories under direct control increased by a factor of nine. As white colonization expanded and systems were established for transportation by land and sea, and as new technologies were introduced, Europe was able to draw upon the resources of the entire planet.

The period between 1850 and 1873 is characterized by an unprecedented economic development, spearheaded by England, which took advantage of its industrial superiority, its colonial empire, and free exchanges. The value of world commerce nearly quadrupled between 1850 and 1880, and the United Kingdom made up the lion's share. In 1880, it held 46% of the world's mercantile tonnage while continually increasing investments abroad.

The British mandate to promote progress and civilization was a pretense to legitimize the imperialism that Benjamin Disraeli and Queen Victoria embodied.

Between the end of the 19th century and the beginning of the 20th, the drive of imperialism and contrasts between powers sharpened; at the center was the race among European states for the division of Africa. In 1905, Russia suffered catastrophic defeat in the Russo-Japanese War, over Korea and Manchuria, the first defeat of a major European power. Imperialism was rooted in economic, political, and ideological causes—including a belief in the superiority of the white race—and enjoyed a broad social consensus, taking advantage of the ties that existed among these motivations. At stake was Europe's domination of the world. The triumphant Age of Imperialism faded with the World War I, which began a long period of economic, ideological, and political conflicts. In the decades that preceded the war, the ranking of industrialized nations had undergone a significant and definitive dislocation: Germany bypassed Great Britain, but the United States, in 1914, boasted an industrial production that was superior to those of the two European powers combined.

SOCIAL AND POLITICAL MOVEMENTS

10. WORKERS AND FARMERS: ORGANIZATIONS, BATTLES, UTOPIAS

The unionization of workers is an important social phenomenon grounded in ancient traditions allied to the trades. These developed in the 19th century in a spontaneous and multifaceted way, even on the international level, taking the forms of societies for mutual assistance, resistance cooperatives for production and consumption, unions, and political parties.

The workers, directly involved in these rapid and menacing changes—at least this was how the changes were perceived—attempted to defend and to organize themselves, to recreate social ties and to establish forms of solidarity. At their core, warlike minorities aimed at radical modifications of the relationships among the classes. This objective was pursued through revolutionary means, or through reformism and gradualism, or it was given over to utopian solutions. The peasants were an integral part of the working class. They were involved directly in the industrialization process. Industrial work was widespread in the countryside in the 19th century; on the other hand, it was the peasants who fed the masses of unskilled laborers whom the manufacturing sector needed in both urban and rural areas. An alliance between laborers and peasants was not possible, partly because the organized workers' movement was deeply divided on political grounds.

The First International Workingmen's Association, assembled in London in 1864 and based on the principal of brotherhood among workers, was characterized by fighting among differing positions. Marx's view favored class struggle and the political party; the view inspired by Proudhon favored cooperation and mutualism; and Bakunin's favored revolutionary anarchism and urged against involvement in politics and insurrection. The differences were accentuated after the defeat of the Commune in Paris in 1871; this brief experiment arose as a symbol of the proletarian revolution but was quashed by a coalition of conservative forces.

The Second International arose of the need for coordination among the various unions and political organizations of the workers' movement that had been structured along national lines. A

1. Mikhail Alexandrovich Bakunin (1814–1876) brought to the European revolutionary movement a Russian radicalism filtered through Hegelian philosophy. An active participant in the 1848–1849 revolution, he was the principal theoretician and organizer of anarchism, which was in conflict and contention with Marx. According to Bakunin, social revolution would be the product of the direct action of the peasants and poor workers, leading to the immediate abolition of the state.

2. Gustave-Rodolphe Boulanger, Clashes Between the Army and the Communards in the Place de la Concorde, *Musée Carnavalet, Paris. The Commune in Paris, March 18 to May 28, 1871, arose in the climate of the fall of Napoleon III, as a symbol of the workers' revolution that was meant to lead to society's self-government. Internal divisions and the isolation from the rest of the country facilitated the Commune's bloody repression at the hands of the troops of Adolphe Thiers.*

3. International Socialist Congress in Zürich, Switzerland, August 12, 1893. The photograph depicts several leaders of European Socialism during a pause in their work. The Congress of Zürich contributed to defining the structure of the Second International, in which the influence of German social democracy was preponderant.

4. Propaganda postcard circulated in Belgium for May Day, 1891. The date of May 1 spread rapidly, on a truly international scale, beyond the expectations of the organizations of the workers' movement. Because of government prohibitions, the holiday was celebrated on the first Sunday of the month of May.

similar congress was held in Paris in 1889, which reached the important decision to proclaim May 1 of 1890 "May Day," an international day of celebration and struggle.

The goal of an eight-hour work day was proposed as the objective of the international workers' movement. The original May Day movement embodied the utopian push toward the unification of the human race. The following congresses witnessed the hard-won victory of the Marxist line, under the guidance of German social democracy, which favored the political parliamentary struggle. The coexistence of reformists and revolutionaries fueled heated debate on such topics as colonialism, militarism, and war. When war manifested itself in Europe with the crisis of July of 1914, it brought to light the prevalence of the national dimension with respect to the ideals of internationalism. With few exceptions, the Socialist parties and the workers' unions, notwithstanding their internationalistic ideologies, all sided to support their homelands.

6. *Giuseppe Pellizza da Volpedo,* Fiumana, *1896. In this painting, the Italian artist (1868–1907) deals with the same topics as those depicted in his more famous* The Fourth Estate *(1901), in which the workers, both farmers and laborers, men and women, express with strength and composure the sense of a march for emancipation.*

5. *Reading,* The British Workman, *London, 1880. A young worker in a blacksmith shop reads a newspaper, which is spread over an anvil, during a pause in his work. The drawing symbolized the workers' capacity for self-emancipation.*

7. *Robert Koehler,* The Strike, *1886. A print of the painting by Koehler served to recreate the atmosphere of events in America in May of 1886. These formed the origin of the May Day tradition as a day of festivities and workers' struggles.*

ATTITUDES AND CULTURES

11. The Victory of Progress

The convictions which dominated the second half of the 19th century were theorized by philosophers like Auguste Comte. These theories included scientific knowledge as the final goal of the history of thought; industrialized society as the necessary and positive result of all human history; Nature as the great reservoir from which the through iron architects and builders could overcome the notion that only structural compactness could guarantee a solid construction.

In the great expositions, not only Progress but also colonialism and military power were exalted. An exhibition organized in Palermo, Sicily in 1891 offered the opportunity to visit an Abyssinian village that had been reconstructed for the occasion. At the World Exhibition in Paris nine years later, the French arms industry flaunted its strength in a menacing-looking pavilion. Most people of the time already accepted the tight relationship between technological-scientific developments and political-military and economic power. The scientific-technological

1. *The construction of the Eiffel Tower on the occasion of the 1889 World Exhibition represented, more than any other enterprise, the conviction in a new possibility of progress.*
2. *The triumph of the new photographic technique, in a recreation of a lithograph by Honoré Daumier from 1860. The celebrated photographer Nadar photographs Paris from a balloon.*
3. *An old woman shows her own image, taken many years before. The photograph not only altered the universe of visual perception, but also it provoked a revolution in the concept of time and in the sense of individual life.*

productive forces can draw infinitely; and the now certain and irreversible affirmation that Progress, assured through science and technology, was destined to extend to every sphere of life and to all of humanity.

The "great expositions" that succeeded each other in various European and American cities in this period were spectacular and splendid celebrations of that optimistic vision. This vision was represented by the construction of several colossal buildings that, with their constructive and architectural concepts, testified to the benefits of technology. The Crystal Palace, from the Great Exhibition in London in 1851, and the Eiffel Tower, from the Exposition in Paris in 1889, provided the proof that

superiority of the West with respect to the rest of the world seemed undeniable. Thinkers of the time, like the Englishman Herbert Spencer, to whom Eurocentrism was not a prejudice but a set of scientific criteria by which to classify human

European expansion hid, according to the novelist Joseph Conrad, a "heart of darkness." The crude reality was that the civilizing mission of the West was simply an "armed robbery" of "those whose skin is different from our own."

Already in 1864, the American George Perkins Marsh, witnessing massive deforestation—of which he himself was a part—denounced the irreparable damage that many human actions had caused to the environment. Also, the certainty of scientific truth was challenged by Friedrich Nietzsche, who in his last works began to abandon the concept that time progresses in a linear and irreversible direction.

In any event, it was not philosophy alone, but also the inventions that rapidly translated into *experience* for enormous numbers of people, that elicited new perceptions of time. Electric lighting appeared to the people of the time an instrument that could nullify the difference between day and night; the editing techniques of the cinema proposed jumps and upturns in time. For several decades, the photograph, offering itself as an unedited substitute to memory, had been altering the sense of the past. The telephone accentuated the sense of simultaneity that the telegraph had awakened, modifying the perception of the present. The present was becoming always more synchronized, regulated by millions of pocket watches—which had evolved from a symbol of social distinction to an instrument of daily life—and governed by laws of punctuality guaranteed by the international treaty signed in 1884 in Washington. The first conference on the fundamental meridian had led to the decision to divide the earth into 24 sections, separated by intervals of one hour, beginning from the zero meridian at Greenwich, England.

Thresher built in Cardiff in 1872, activated by a locomobile *or portable steam engine, a self-propelled steam engine with various industrial and agricultural applications, including plowing, reaping, and hoeing. In use in the United States since the beginning of the 19th century, the portable steam engine first appeared in Europe at the Great Exhibition of London in 1851.*

INDEX

Each entry is followed by the number(s) of the chapter(s) in which it appears.

Adler, Dankmar 6
agrarian industrialism 1
agriculture 1
Albert (Queen Victoria's husband) 9
Alberta 1
Algeria 9
alternators 5
American Telephone and Telegraph Company, the (AT&T) 8
Ampère, André-Marie 5
anarchism 10
anesthesia 2
anesthetics 2
Anshutz, Thomas 6
anti-rabies vaccine 2
antibodies 2
arc lamp 5
Argentina 1
aristocracy among workers 6
Arkwright, James 4
arms industry 11
artisan workshop 4
Atlantic Ocean 7
Australia 1
automatic switching 8

bacteria 2
Badger, Daniel D. 6
Bakunin, Mikhail Alexandrovich 10
Belgium 1
Bell, Alexander Graham 8
Belle Époque: introduction
Berlin 2, 3, 8
Bessemer, Henry 6
Bessemer's convertor 6
Bismarck, Otto von 8
blast furnaces 6
blood pressure machine 2
boring machine 4
Boulanger, Gustave-Rodolphe 10
Boxers 9
Brazil 1
Brunel, Isambard Kingdom 7

cacao 1
California 3
camera 11
Canada 1
Canadian Pacific Railroad 1
Carnot, Sadi 7
cast iron 6
cattle farm 1
chemical fertilizers 4
chemistry: introduction, 4
Chicago 8
child labor 6
child mortality 2
China 9
chloroform 2
cholera 2
cinema 11
city(ies): introduction, 2, 3, 8
Civil War 1, 9
clocks 11
coal 5, 6, 7
coffee 1
coke 6
collective mentality 11
colonialism 9, 10, 11
colonization 1, 9
Commune, the (Paris, March 18–28, 1871) 10
Comte, August 11
Congo 9
Conrad, Joseph 7, 11
cooperation 10
cooperatives for production 10
cotton 1

Crompton, Samuel 4
Crystal Palace 11
Cuba 1
cultures 11
cylindrical grinders 1

daily life 2, 5
Daumier, Honoré 11
Davy, Humphry 5
deforestation 11
Degas, Edgar 9
department stores 3
Diesel, Rudolf 4
disinfection 2
Disraeli, Benjamin 9
doctors 2
durum wheat 1
dynamo 5

East India Company 9
École Polytechnique 4
Edison, Thomas Alva 5
Egypt 9
Eiffel, Alexandre-Gustave 6
Eiffel Tower 11
electric lighting 11
electric shutters 11
electric plant 5
electricity: introduction, 4, 5
electromagnet 5
emigration 3
empire: introduction
energy: introduction, 4, 5, 7
England 4, 6, 9
entrepreneurs: introduction
environment(s) 3, 6, 11
epidemic 2
ether 2
Eurocentrism 11
Europe 2, 3, 7, 8, 9, 10

factories 4, 5. 6
factory-cities 3
Faraday, Michael 5
farmers 10
First International Workingmen's Association 10
forests 11
Franz Joseph, Emperor of Austria 8
fundamental meridian 11

gas 5
gas lighting 5, 6
gastroenteritic maladies 2
German social-democracy 10
Germany 4, 6, 9
germs 2
Giessen 4
gradualism 10
great expositions 11
Great Britain 2, 4, 9
Great Eastern 7
Great Exhibition (London, 1851) 11
"Great Revolt" of 1857–1858 (India) 9
Greenwich 11
grinding 4
Gualdry, Ferdinand 4

Hapsburg 8
Haussmann, Georges-Eugène 2, 3
Hargreaves, James 4
Hinckley, Robert 2
Hofman, von 4
home 3, 5, 6, 8
horse-drawn omnibuses 2
housewives 8
hygiene 2

30

4. In 1878, Muybridge aligned 12 cameras with electric shutters timed at a thousandth of a second, about 27" (70 cm) apart, activated as a racehorse broke a series of twelve strings pulled taut across a track. The frames showed the anatomy of a running horse, which had never been seen by zoologists or artists—indeed, by the human eye. In 1880, Muybridge created a machine, called the "zoopraxiscope," that rapidly turned the photographs of the horse on a circular pane of glass, projecting onto a screen the illusion of motion.

5. A brochure published for the first Kodak camera in 1888. Photography rapidly came into use by the masses and gave rise to a new and large industry. In 1889, the American company Kodak advertised its cameras with the slogan "You push the button, we do the rest."

6. The gunboat-shaped pavilion of the Schneider Company from Le Creusot, constructed for the Parisian World Exhibition of 1900.

societies and stabilize the hierarchies among them, undoubtedly contributed to the notion that progress and competition had become synonymous. In the first half of the 20th century, and even in the second half of the 19th century, dissonant voices were expressed that countered the collective mentality and the theories that dominated world culture. The triumphal